D1551850

ECOSYSTEMS

Life in a Coral Reef

Hayley Mitchell Haugen

KIDHAVEN PRESS™

San Diego • Detroit • New York • San Francisco • Cleveland
New Haven, Conn. • Waterville, Maine • London • Munich

Picture Credits

Cover: PhotoDisc
© Australian Picture Library/CORBIS, 19
© Hal Beral/CORBIS, 30
© Ralph Clavenger/CORBIS, 27
© COREL Corporation, 7, 15, 17, 22, 29, 31
Florida Keys National Marine Sanctuary/NOAA, 5, 11, 13
© Stephen Frink/CORBIS, 24, 34
Dr. James McVey/NOAA, 15 (inset)
Mr. Ben Miermet/NOAA, 18
Mohammed Al Momany/NOAA, 6
Brandy Noon, 12, 28, 33
NOAA, 25, 36, 40
PhotoDisc, 13 (inset)
© Steve Rayner/CORBIS, 38
Martha Schierholz, 9

© 2003 by KidHaven Press. KidHaven Press is an imprint of The Gale Group, Inc., a division of Thomson Learning, Inc.

KidHaven™ and Thomson Learning™ are trademarks used herein under license.

For more information, contact
KidHaven Press
27500 Drake Rd.
Farmington Hills, MI 48331-3535
Or you can visit our Internet site at http://www.gale.com

ALL RIGHTS RESERVED.
No part of this work covered by the copyright hereon may be reproduced or used in any form or by any means—graphic, electronic, or mechanical, including photocopying, recording, taping, Web distribution, or information storage retrieval systems—without the written permission of the publisher.

LIBRARY OF CONGRESS CATALOGING-IN-PUBLICATION DATA

Haugen, Hayley Mitchell, 1968–
  Life in a coral reef / by Hayley Mitchell Haugen.
    p. cm. — (The ecosystems library)
  Includes bibliographical references (p.).
    ISBN 0-7377-1370-4 (hardback : alk. paper)
    1. Coral reef ecology—Juvenile literature. 2. Coral reefs and islands—Juvenile literature. [1. Coral reefs and islands. 2. Coral reef ecology. 3. Ecology.] I. Title. II. Series.
    QH541.5.C7 H38 2003
    577.7'89—dc21
                                                         2002010666

Printed in China

Contents

Introduction
The Coral Reef Ecosystem — 4

Chapter 1
What Is a Coral Reef? — 10

Chapter 2
Other Animals in a Coral Reef — 21

Chapter 3
Endangered Coral Reefs — 32

Glossary — 42
For Further Exploration — 44
Index — 47

Introduction

The Coral Reef Ecosystem

Coral reefs cover 80 million square miles of Earth's shallow, tropical waters. The reefs are prized by marine biologists, ecologists, fishermen, and travelers. Often referred to as the rain forests of the sea, coral reefs are some of the most complex **ecosystems** on Earth. They support more than five thousand species of coral and up to three thousand species of fish and **invertebrates**.

In this environment of corals, sea cucumbers, sponges, clams, tropical fish, and thousands of other marine species, the coral reefs are awash in all the colors of the rainbow. In the coral reef is a world of wondrous shapes and textures and unending movement. Marine creatures creep, swim, and glide, and plants bend, float, and sway to the rhythm of the sea.

Most of the marine life in this underwater world feed on plants and other sea creatures for survival. Some coral reef organisms work together to protect one another from harm. This type of relationship is called **symbiotic**. The clown fish and the sea anemone are good examples of symbiosis.

The color of the bright orange clown fish makes it an easy target for predators. However, the clown fish, which is covered with a sticky **mucus** that protects it

Coral reefs, often referred to as the rain forests of the sea, are home to thousands of plant and animal species.

A clown fish hides from predators among the stinging tentacles of a sea anemone.

from the sea anemone's stinging **tentacles**, is able to hide among the tentacles of sea anemones. Meanwhile, the anemone's tentacles are seen more clearly against the brightly colored clown fish and scare away potential predators.

Algae in the Coral Reef

Marine animals bring both color and excitement to the coral reef, but they are not alone in this environment. The corals, fish, and other inhabitants of coral reefs share their habitat with an abundance of plant life as well. In fact, sea creatures in the coral reef could not survive without the aid of marine plants. Even though coral reef plants are not as varied as the animals found there, their role is important for the success of the coral reef community.

The most abundant marine plants in coral reefs are seaweeds. Seaweeds are mostly made up of different species of **algae** that are especially important for helping the reef to function. Algae are diverse but simple, plantlike organisms. They are separated into three groups by color: greens, reds, and browns. Like coral, algae come in all shapes and sizes. Turtle weed, for example, is a green algae that grows in hairy tufts.

Red, green, and brown algae play an important role in the coral reef community, offering food and shelter to many sea creatures.

The largest forms of algae are seaweeds that stretch up to three hundred feet from the ocean's floor to the water's surface. Algae do not live only in the ocean, however. Algae are able to survive in almost any habitat that has plenty of sunlight and moisture. They can live in coral reefs, freshwater lakes, ponds, streams, swamps, hot springs, and even Arctic glaciers.

Unlike plants, algae have no roots, stems, or leaves. Like plants, though, most algae use the sun's energy through **photosynthesis** to make their own food. In fact, because they gather up more of the sun's energy and produce more oxygen than all other plants combined, algae are considered the most important photosynthesizing organisms on the planet.

In some species of algae, such as the red variety called coralline algae, cell walls become hardened with **calcium carbonate**. Calcium carbonate is the substance that makes up the tiny skeletons of coral **polyps**. The rigid cell walls of this algae allow it to attach itself to solid surfaces. Red coralline algae attaches to the ocean floor and calcium carbonate begins to form. It then cements itself to other organisms, providing the coral reef with new material for growth.

Studying Coral Reefs

Because coral reefs support such a wide array of marine plants and animals, they are fascinating for scientific study. Scientists also study and appreciate coral

This octopus is one of the many fascinating creatures that live in coral reefs.

reefs' benefits to humans. Geologists interested in fossil fuels, for example, often study the petroleum and gas deposits found within the oldest reefs. Coral reefs also benefit the world of medicine. More than twelve hundred biochemical compounds discovered in coral reef organisms are currently used in the treatment of asthma, arthritis, cancer, AIDS, and other ailments.

Chapter 1

What Is a Coral Reef?

Coral often makes people imagine something as hard as rock and as colorful as a seashell found along the beach near their home. These images are not incorrect. Many corals *are* hard to the touch and sharp, too. And corals *do* come in as many dazzling colors as the tropical fish that make coral reefs their home. But corals are more than just rocky, vibrant exteriors. The world of corals is exciting and full of surprises.

Corals Are Animals

Although some corals feel as hard as rocks, they are not rocks at all. They are simple but delicate animals, and their more formal name is coral polyps. Each coral polyp has a hollow body in the shape of a tube, like a small balloon, and a tiny skeleton made up of calcium carbonate.

At the open end of its body is the coral's mouth. It is equipped with tiny poisonous tentacles used for stinging prey and protecting itself from predators. In this manner, corals are like jellyfish or sea anemones. This is not surprising because many corals, such as the grape coral, are related to the jellyfish and sea anemone family.

Unlike jellyfish though, corals remain in one place on the reef and do not swim in the water. A

A French angelfish (right) makes its home among the colorful corals of a reef.

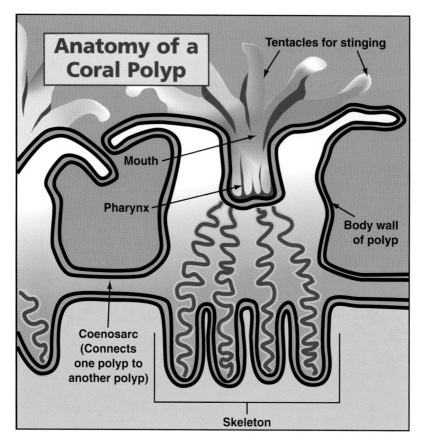

coral's tiny tentacles can sway back and forth in the water, but the animal relies on the water's motion to eat. The movement of the seawater pushes the food that corals eat into their tentacles. These creatures are called **zooplankton** and include tiny newly hatched shrimp, crabs, sea worms, and other creatures too small to see with the human eye. First, the zooplankton are stung by the coral polyp's tentacles. Then, tentacles move the zooplankton into the polyp's mouth for digestion.

Even though the water's movement can provide coral polyps with zooplankton twenty-four hours a day, coral polyps feed only at night. During the day they hide from predators by retreating into the stony shell they have created for themselves. These stony shells make up the coral reef. The protective shell that surrounds each coral polyp is made of a material called corallite.

Feasting on zooplankton at night does not provide coral polyps with enough food for survival. The **stony corals**, or hard corals, that live near the water's surface also rely on an algae called **zooxanthellae** for survival. This algae lives in the cells of the coral polyps' tiny bodies. Zooxanthellae captures energy from the sun through the process of photosynthesis.

Tiny stinging tentacles allow coral polyps to catch and eat zooplankton (inset).

It then shares this energy with coral polyps and receives protection from plant-eating predators in exchange. With plenty of zooplankton to eat and zooxanthellae creating additional energy for coral polyps, the coral reef can continue to grow.

The Coral Colony

Coral reefs do not grow quickly, however. It actually takes between five thousand and ten thousand years for coral reefs to develop. And in addition to needing plenty of food to eat, the coral reef habitat also needs the right environment to encourage reef growth. Coral reefs, for example, need clear, shallow, salty seawater that reaches temperatures between 75 and 85 degrees Fahrenheit. It is only in these conditions that a coral colony can form.

The coral colony is made up of millions of corals clustered together. When the corals grow individually, they are similar to plants because they form buds that branch off each coral and grow upward like trees. In addition to budding like plants, coral polyps also produce eggs. When the eggs are fertilized, they detach from the coral polyp and create new coral reefs atop rocks or dead coral colonies.

Corals attach themselves to these rocks or colonies with the aid of a limestone cup that gradually forms to surround and protect the polyp's body. Cells outside the coral polyp collect calcium carbonate from the seawater to form this limestone cup. When the coral polyps die, these hard shells remain.

A scuba diver explores a sea cave lined with sea fans and soft corals. At right is a reef-building hard coral.

The reef grows as new coral polyps attach to the old shells.

Although many kinds of corals exist, only stony corals that contain a calcium carbonate skeleton can help build reefs. These hard corals are collectively known as reef-building corals. They can be found along the eastern shores of Australia and Indonesia, around the Caribbean islands, where more than sixty species of coral reside, and in many tropical waters around the globe.

Soft Corals

Because of their relationship with zooxanthellae, the hard reef-building corals need to live close to sunlight on the water's surface to survive. But sea whips, sea fans, and other soft corals that do not house zooxanthellae can live in dark caves in the ocean or even deep on the ocean floor itself. Soft corals are often the most colorful corals. Corals such as the vibrant orange daisy coral contain bright orange polyps surrounded by spectacular orange-yellow tentacles. This species of coral lives among shipwrecks and in caves some thirty-three to fifty feet below the water's surface.

Unlike reef-building coral, soft corals are not sheltered by a limestone cup. Instead, they are protected by **spicules**. Spicules are tiny needles of limestone that can bend and sway in the water. The flexible internal skeletons of soft corals, such as sea fans and sea whips, allow them to bend without breaking as they hunt food in the sway of the ocean. Like stony coral, soft corals eat zooplankton to survive, but they do not need to host plants within or on themselves for further nourishment.

Three Kinds of Reefs

Just as different kinds of corals live in the reef environment, various types of reefs exist. A common variety of reef is the **fringe reef**. This reef forms along the border, or the fringe, of a shoreline. A fringe reef might, for example, grow out from the rocky ledge of an island that connects with the sea. Because fringe reefs are separated from the land by only a small area of water, they are usually shallow enough for people to wade across and explore during low tide.

The most unusual kind of reef is the coral **atoll**. These reefs are usually found in very deep water in the Indian and Pacific oceans. Not attached to the land in any way, an atoll is a ring of coral that seems to float just under the surface of the water.

More than one hundred years ago scientist Charles Darwin studied the formation of these unique reefs. He discovered that atolls begin as fringe reefs surrounding volcanic islands. When the ocean floor shifts over time, however, these volcanic

A juvenile snapper finds refuge among the arms of a soft coral.

Charles Darwin proposed that coral atolls begin as fringe reefs and develop, with time, into beautiful coral islands like this one.

islands eventually sink into the ocean and leave only the ring of coral behind.

Over time, coral atolls can develop into coral islands up to twenty feet high. These small islands, such as the Maldive Islands in the Indian Ocean, are formed when sand begins to build up and eventually cover the reef. Plants and trees start to grow from this sand once seeds begin to take root there.

The third and most common type of reef is the **barrier reef**. Similar to fringe reefs, these reefs also grow parallel to, or next to, the land. Unlike fringe reefs, however, they are separated from the shore by a large area of calm water. This area between the shore and the reef is called a **lagoon**. And on the other side of the lagoon and the reef is the far-reaching expanse of the ocean.

The Great Barrier Reef

As its name suggests, the Great Barrier Reef off the northeastern coast of Australia is by far the largest reef in the world. Scientists speculate that the Great Barrier Reef began forming 30 million years ago during an ice age. It stretches more than 1,260 miles, with some parts reaching more than 100 miles across the ocean from the coast. This reef is so large that it can be seen from space!

A scuba diver exchanges glances with a giant codfish on Australia's Great Barrier Reef.

Considering its great size, it is not surprising that this reef is made up of a number of smaller reefs that are separated by channels of water. Nevertheless, it is the largest-ever structure in the world built by animals. One positive effect of this great structure is that it acts as a strong wall-like barrier that protects Australia's coastline against the crashing waves of the Pacific Ocean.

The Great Barrier Reef also serves as home to about four hundred different species of coral, fourteen hundred species of fish, and more than two thousand other marine species such as eels and octopus. Australia's Great Barrier Reef is unique because it contains such a large number of coral, fish, and other marine species. But similar animal species can be found in coral reefs throughout the world.

Chapter 2

Other Animals in a Coral Reef

With as many as one hundred and fifty thousand different animal species living in some of the largest coral reefs, it is easy to see why people often think of coral reefs as busy centers of ecological activity. In addition to the colorful variety of corals, anemones, **crustaceans**, eels, and sea slugs, a single coral reef can contain up to five hundred species of fish. Although the coral reef might easily become overcrowded under these conditions, the reef inhabitants have created an inventive way of preventing this problem.

The marine animals live their lives in the reef on different schedules. Up to two-thirds of the colorful reef fish, such as angelfish, for example, are diurnal,

A stealthy reef shark comes out at dusk in search of its next meal.

meaning they are most active during the day. The other one-third of the reef fish, such as squirrelfish, as well as many invertebrates, such as crabs, lobsters, and sea urchins, remain out of sight during the day and take over the reef at night. Like land animals that are active at night, these fish are known as nocturnal. The reef schedule also includes special in-between hours for about 10 percent of the reef species that are crepuscular, meaning most active at dusk and dawn. These species include most of the predators, such as reef sharks and barracuda.

Awash in a Sea of Color

The vibrant and exotic colors of coral reef fish capture the imagination of nature enthusiasts worldwide. Elsewhere in nature, such as in a rain forest or on an African savanna, the animals have adapted their plumes and coats to blend in with the scenery to protect themselves from predators. In the coral reef, however, the various species boldly express, rather than hide, their bright and often unusual markings.

Surprisingly, the coral reef's colorful species are actually using their bold colors to protect themselves. Marine biologists have discovered that most ocean species view the world in ultraviolet and polarized light. This means that they can see many aspects of light that land animals, including humans, cannot. The bright colors and markings on the coral reef animals help them reflect ultraviolet light as a means of communicating with each other.

The bright yellow and blue stripes of the easily recognized emperor angelfish help members of this species recognize each other, even in dim light. When mating season begins, this fish's colors can become even brighter to help it attract a mate. Other reef species use their colors to show aggression and warn others, or to simply fade into other colorful spots of the reef to avoid predators.

Colorful marine animals can even signal each other by spreading their fins in order to flash special markings

The emperor angelfish uses its vibrant blue and yellow patterns to attract a mate, warn predators, and for camouflage.

that reflect ultraviolet light. Some fish may flash these normally concealed areas to warn predators away, while others may wish to draw attention to themselves. The mandarin fish is just one coral reef fish whose colors offer it protection. This bright, intricately designed fish has four colors: orange, green, blue, and yellow. The mixture of bright colors and swirling patterns warn other fish that the mandarin fish is covered in a bad-tasting, slimy mucus. For most predators, the fish's colorful appearance is enough to warn them away.

Crustaceans Among the Coral

In addition to the hundreds of species of tropical fish that inhabit the coral reef, many varieties of crustaceans live there as well. Crustaceans, such as crabs, lobsters, shrimp, and barnacles, have bodies covered with a hard shell or crust. One common coral reef crustacean is the hermit crab.

Hermit crabs take up residence in the discarded shells of whelks and sea snails. When hermit crabs outgrow one shell, they search for another, larger home. On the reef, some hermit crabs will even choose to live inside tubes in the coral itself. With its eight legs and a large pair of pincers, the hermit crab has no problem catching food or protecting itself from predators.

A lobster seeks cover among the reef's countless cracks and crevices.

Another crustacean found in coral reef habitats is the reef lobster. Smaller than red lobsters sold for consumption in grocery stores, the reef lobster grows up to only five inches long. Its white body, bright lavender spots, and hot pink pincers with orange-ellow tips also distinguish this reef dweller from other lobster species. Rarely seen out in the open, this lobster hides in dark caves and crevices within the coral reefs of Indonesia.

The Sponge

A coral reef species that may offer housing for hermit crabs, reef lobsters, and a host of other small reef animals inside its body cavity is the sponge. Although they are the most simple of all multicellular organisms, sponges, just like corals, are animals that attach themselves to the reef. Sponges are found in abundance in tropical waters, where they help corals in the reef-building process.

The ten thousand different species of sponges exist in a wide variety of shapes, sizes, and colors. The pink tube sponge is appropriately named because this sponge grows in clusters of pink tubes. Each tube can grow from eight to sixteen inches long. This sponge is often covered with wormy sea cucumbers, which help clean the sponge of debris by eating the various organisms that have attached to its sides.

Starfish: Friend and Foe

Also contributing to the colorful landscape of the coral reef are starfish. Starfish, or sea stars, are found

A cluster of tube sponges extends out from a coral forest. Sponges aid coral in the reef-building process.

in every depth of the ocean. Of the fifteen hundred species of starfish, most have stiff bodies with four to six arms radiating from the center. Each of these arms is equipped with rows of suction cups for crawling along and attaching to the reef. Although the starfish moves more by its sense of touch than by sight, at the tip of each arm is an eyespot that is sensitive to light.

Most starfish benefit the coral reef because they feed on debris and keep the populations of other species, such as the burrowing mollusk, in check. But

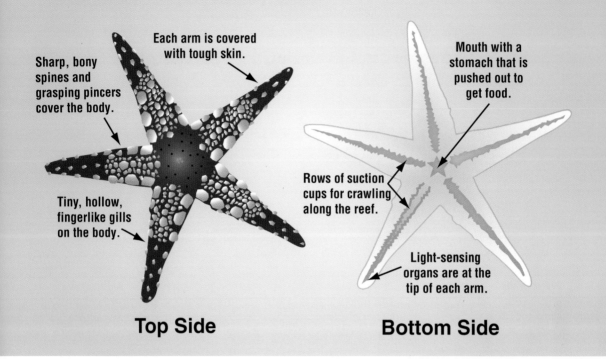

other starfish wreak havoc on the reef environment. Some destroy oyster beds. The crown-of-thorns starfish has the worst reputation of all. It feeds on the coral itself. The beautiful, bright purple and thorny crown-of-thorns starfish appears regularly in coral reefs in Southeast Asia. Feeding off live coral, this starfish consumes large areas of the reefbed, leaving nothing but bleached, dead coral behind.

Sea Cucumber and Urchins

The sea cucumber, related to the starfish, is a marine animal without a head. Instead, it has a soft, rubbery body with a mouth at one end for eating and an opening at the other for expelling waste. Because of their bendable exteriors, sea cucumbers can change shape easily as they creep along the reef with their numerous, brightly colored sucker feet.

The more than seven hundred species of sea urchins are also related to sea cucumbers and starfish. Sea urchins are spiny-skinned animals that use their sharp spines both for movement and for defense. The sharp spines of some species, such as the spiny star urchin, are poisonous to both animals and humans and can cause a painful wound if stepped on.

The sharp spines of this pencil urchin protect the animal from predators.

A Giant in the Reef

Coral reef species such as the starfish are fascinating for their bright colors and endless variety. The reef also houses marine species that are interesting because of their size. One such specimen of the reef is the giant clam. It can measure more than five feet wide and, including its shell, weigh up to five hundred pounds. Giant clams are found at great depths throughout the western Pacific, including Southeast Asia, Australia, and New Guinea. They can live for more than one hundred years.

The giant clam, along with the other various reef species, make up just a small sampling of the wondrous animals found in the coral reef. All of these

This giant clam was discovered near New Guinea. It is just one of many fantastic animals that make their home in the coral reef.

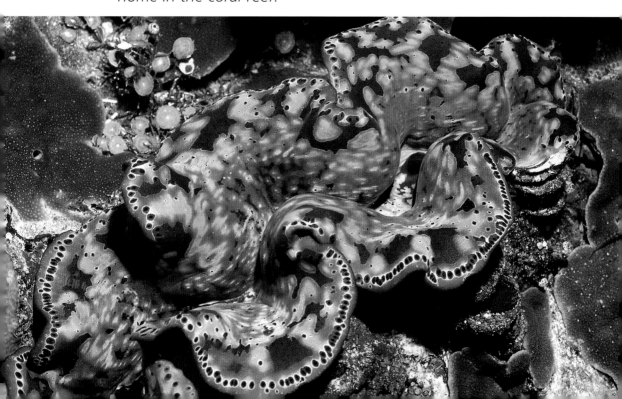

Reef-dwelling animals like this banded sea snake depend on the reef for protection and food.

creatures support the ecology of the coral reef, making room for each other by sharing the reef in day and night shifts, and by nourishing each other as both predator and prey. While the animals support each other, their lifestyles are also protected by their environment. The reef itself provides both protection and nourishment, and in addition to corals, marine plants also play a vital role in the lives of reef animals.

Chapter 3

Endangered Coral Reefs

Coral reefs are in danger. Pollution, global warming, and overfishing are just a few factors causing the destruction of coral reefs worldwide. Scientists warn that unless these destructive forces are controlled, coral reefs will die. At the current rate of destruction, up to 57 percent of coral reefs will die within our own lifetimes.

Coral Reefs Benefit Marine Life and Humans

Coral reefs offer benefits to both marine life and humans. Millions of marine species such as fish, crustaceans, sponges, worms, anemones, and countless varieties of plants rely on coral reefs for both food and shelter. When the reef habitat dies, so do many of these plant and animal species.

Plants and animals are not the only ones harmed by coral reef destruction. Humans suffer when reefs die as well. First, coral reefs help maintain coastal geography by working as natural barriers to protect beaches from erosion and storms. In addition to protecting the land, coral reefs also help people in other ways.

More than 500 million people living in coastal areas in Asia, Africa, and the Caribbean rely on coral reefs for food. A thriving coral reef can produce fifteen tons of fish and seafood for human consumption each year. That is enough food to feed twenty-five hundred people.

Coral reefs are also tourist attractions. In the Virgin Islands alone, more than forty thousand tourists

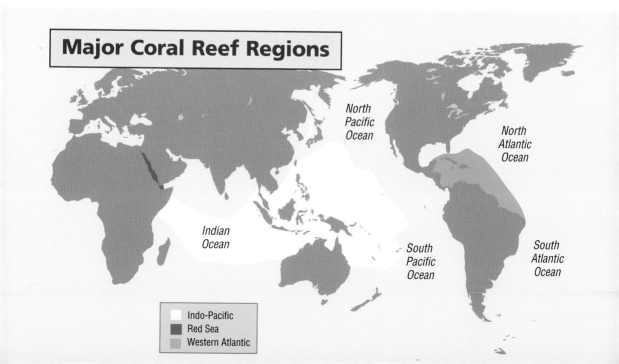

come each year to snorkel or scuba dive in the coral reefs. More than half the people who live on these islands rely on the tourist trade for employment. If the coral reefs die in these areas, many island people will lose their only source of income.

Coral Reefs in Danger

Fisheries and tourism benefit from coral reef habitats, but they can also endanger reefs. Gas and oil from tourist boats can pollute reef waters. Boat anchors can also damage reefs by scraping against the

Even careful snorkelers and scuba divers can damage fragile coral reef ecosystems.

delicate coral and shattering whole areas of reef in a single anchoring.

Divers and snorkelers may also harm reefs. The heavy, metal oxygen tanks that divers use often scrape against the coral, breaking it away. Although the practice is illegal, some divers also break pieces of coral off the reef to keep as souvenirs. Human trash dumped into reef waters also affects the reef's ability to thrive. Trash can get stuck in coral polyps and prevent them from expanding their tentacles to eat.

Pollution and Global Warming Lead to Coral Bleaching

Other types of pollution also affect the reef. Coastal runoff, for example, often finds its way to the coral reefs. The materials found in coastal runoff include human sewage, fertilizers, gasoline, oil, and other man-made chemicals. All of these materials can trigger the overproduction of the algae blooms that smother the coral and lead to **coral bleaching**.

When coral is smothered by algae blooms, it expels the zooxanthellae within the cells of the coral polyps that it needs for energy. Soon after this happens, the coral turns white.

Coral bleaching is a concern for scientists eager to protect coral reefs because coral reefs can take many years to recover from bleaching incidents. With the zooxanthellae expelled, the corals can no longer

Water pollution from coastal runoff causes corals to become bleached like the coral in this photo. Eventually, the corals die and wear away.

make deposits of calcium carbonate, which serve as the building blocks for the reef. Without these deposits, the reef cannot grow.

Eventually, the corals affected by the bleaching die and erode away, so that all the reef's creatures lose the protection of the reef habitat. Scientists are concerned about the effects of coral bleaching because entire colonies of coral can be wiped out by it.

Scientists believe that another cause of coral bleaching is global warming. They suspect that the rising temperature of Earth is to blame for the death of 10 percent of the world's reefs in 1999. Corals are so sensitive to temperature changes in the water that even prolonged temperatures of only four degrees above normal can cause them to bleach and die. Without protection from global warming, some scientists warn, up to three-quarters of the world's reefs could die within the next fifty years.

Overfishing Threatens Coral Reefs

Tourists, pollutants, and global warming are not the only causes of reef destruction. The fishing industry is also responsible for much reef habitat loss each year.

Nearly twenty-five thousand metric tons of reef fish are caught each year for human consumption or for aquarium use. This adds up to big business, about $1 billion a year. With so much money at stake, coral reefs are often overfished so that some fish species do not have a chance to reproduce quickly enough.

Fishing for one kind of fish over another can also disrupt the delicate ecosystem of the coral reef, where marine life relies on the food chain for survival. In Malaysia, Vietnam, Indonesia, and the Philippines, for example, overfishing has allowed algae to grow more quickly than normal because there are not enough fish left in the water to eat the algae and keep its growth under control. Too much algae

Fishing boats fill a harbor in Vietnam. Overfishing is a major cause of coral reef destruction.

in a coral reef can kill the coral. As the coral dies, other species that rely on the coral for their habitat also begin to disappear.

Destructive methods of fishing are another threat to coral reefs. At least ten countries in the world use cyanide, an extremely poisonous chemical, to capture coral reef fish for the aquarium market. To use the cyanide, fishermen fill plastic squirt bottles with crushed cyanide tablets. While underwater, divers simply squirt the cyanide at their prey to stun it.

Although the cyanide stuns some fish, it often kills other marine species, including coral polyps. Cyanide fishing is also wasteful. Many fish collected for aquariums die before they can be sold.

Efforts to Protect and Conserve Coral Reefs

Many countries worldwide have taken steps to conserve coral reef habitats. Mozambique has banned the trading of coral skeletons and stony corals, for example, and the Philippines has created a total ban on coral trading. In addition, the Philippine government is also encouraging fishermen to stop cyanide fishing.

The government of Australia has chosen to zone the country's coral reef for various uses. Some areas may be fished, but other spots are left alone to preserve the stability of the reefs. All coral collectors in Australia must now also carry a special license, and they are allowed to take coral from less than only 1 percent of the reefs in a designated area.

While countries around the globe are doing their share to protect coral reefs, the United States has become the leader in coral reef conservation. One example of reef conservation in the United States is found in the Florida Keys. In 1990, Congress set aside a twenty-eight-hundred square-nautical-mile area known as the Florida Keys National Marine Sanctuary to protect coral reefs in the Keys. In this area, reef and channel markers are used to keep boats out of the conservation area, where fishing and coral collecting are prohibited.

Looking Ahead

The United States has also announced plans to set aside 20 percent of all its coral reefs as conservation sites similar to the one in Florida. The U.S. Coral Reef Task Force hopes to achieve this goal by 2010.

A marine biologist collects coral samples from a protected Florida reef.

The task force has also proposed that the United States spend $25 million each year to continue its coral reef research. All these efforts are fighting to ensure that future generations will be able to experience the beauty and ecological diversity of America's coral reefs.

Glossary

algae: Tiny water plants that lack roots, stems, and leaves, and they provide food for sea creatures.

atoll: A group of islands and coral reefs surrounding a lagoon.

barrier reef: A coral reef near a major shore, usually along the edge of an ocean shelf and separated from the land by a lagoon.

calcium carbonate: A chemical compound from which limestone is formed; also found in marble, coral skeletons, crabs, shells, teeth, and bones.

coral bleaching: When corals turn white and eventually die from the loss of zooxanthellae algae from within the polyps.

crustaceans: A group of animals with jointed legs and a shell, but no bones, such as lobsters, crabs, and shrimp.

ecosystems: Living species and their habitats within a certain region.

fringe reef: A coral reef running parallel to the shore.

invertebrates: Animals without backbones.

lagoon: A shallow pond or water channel near or connected to a larger body of water.

mucus: A slimy, often poisonous substance that some coral reef animals produce.

photosynthesis: The process by which plants convert light into energy.

polyps: Coral or other soft-bodied animals with a tubelike body and a ring of tentacles around the mouth.

spicules: Tiny needles of limestone that strengthen the bodies of soft corals.

stony corals: Corals that produce external limestone skeletons.

symbiotic: A beneficial relationship between two species.

tentacles: Flexible feelers for touching, feeding, and smelling.

zooplankton: Tiny creatures that float through the water and are eaten by coral polyps.

zooxanthellae: One-celled algae living in the tissues of reef-building corals, providing them nourishment.

For Further Exploration

Books

Gerald R. Allen, *Tropical Marine Life*. North Clarendon, VT: Periplus Editions, 1997. This nature guide provides an introduction to the marine plants and invertebrates that inhabit the coral reef ecosystems of the Indo-Pacific region.

Norman Barrett, *Coral Reef*. New York: Franklin Watts, 1991. This Picture Library reference book uses color photography and simple text to illustrate the varieties of life in the coral reef.

Dirk Bryant, *Reefs at Risk*. Washington, DC: World Resources Institute, 1998. For more advanced readers, this book discusses the relationship between coral reefs and people, threats to the world's coral reefs, and what is being done to save them.

Mary M. Cerullo, *Coral Reef: A City That Never Sleeps*. New York: Dutton, 1996. Written by an aquarium educator, this book describes the colorful daytime feeders of the coral reef and the blind nighttime inhabitants that rely on their senses of smell, taste, and touch as they maneuver through the reef at night.

Sylvia A. Earle, *Hello Fish: Visiting the Coral Reef*. Washington, DC: National Geographic Society,

1999. A noted marine biologist and ocean explorer takes readers on a tour of coral reefs around the world in this colorful, fact-filled book.

Rebecca L. Johnson, *The Great Barrier Reef: A Living Laboratory*. Minneapolis: Lerner Publications, 1991. Accompanied by many color photographs, the text of this book offers a firsthand account of scientists' research projects in the Great Barrier Reef.

Caitlin and Thane Maynard, *Rain Forests & Reefs*. New York: Franklin Watts, 1996. In this book, photographs, journal entries, and postcards document the experiences of the Junior Zoologists club as they learn about endangered species of the rain forests and coral reefs in Belize, Central America.

Laurence Pringle, *Coral Reefs: Earth's Undersea Treasures*. New York: Simon and Schuster, 1995. An award-winning science writer describes the relationships among fish, crabs, and other coral reef animals. The increasing threats to coral reefs are also discussed.

Barbara Taylor, *Coral Reef*. New York: Dorling Kindersley, 1992. From the Look Closer series, this book gives readers a close-up look at crabs, fish, anemones, and other coral reef wildlife in their natural surroundings.

Jenny Wood, *Coral Reefs: Hidden Colonies of the Sea*. Milwaukee: Gareth Stevens Children's Books, 1991. The mysteries of coral reefs, fringe reefs, and barrier reefs are explored. Also included is a true-life story of one family's shipwreck adventure.

Websites

Coral Reef (Visible Earth) (http://visibleearth.nasa.gov). The Visible Earth project from NASA provides satellite pictures of coral reefs.

ReefNet (www.reefnet.org). This site offers resources for coral reef and marine conservation.

Slash Zone: Rock and Reef Homes (www.mbayaq.org). Sponsored by the Monterey Bay Aquarium, this online exhibit features rocky coast and coral reef animals. Includes games and audio features.

Index

algae
  blooms of, 35–36
  calcium carbonate and, 8, 36
  overfishing and, 37–38
  photosynthesis and, 8, 13–14
  types of, 7, 8, 13
angelfish, 23
atolls, 17–18
Australia
  conservation in, 39
  giant clams in, 30
  Great Barrier Reef of, 19–20

barracuda, 22
barrier reefs, 18–20
benefits, 9, 32–34

calcium carbonate
  algae blooms and, 36
  limestone cups and, 14
  skeleton of coral and, 8, 11
clams, 30
clownfish, 5–6
coastal runoff, 35–36
colors
  of algae, 7
  of crustaceans, 26
  for protection, 23–24
  of soft corals, 16
  of sponges, 26
  of starfish, 28
conservation, 39–41
coral bleaching, 35–37
coralline algae, 8

corallite, 13
coral polyps
  algae in cells of, 13–14
  bodies of, 8, 10–11
  cyanide and, 39
  diet of, 12–13
coral reefs
  environment needed by, 14, 37
  extent of, 4
  growth process of, 14–15
  location of, 15
  number of species of coral in, 4
  types of, 16–19
crabs, 25
crepuscular animals, 22
crown-of-thorns starfish, 28
crustaceans, 25–26
cyanide, 38–39

Darwin, Charles, 17–18
destruction
  by boat damage, 34–35
  by coastal runoff, 35–36
  by cyanide, 38–39
  by global warming, 36–37
  by overfishing, 37–38
  by tourism, 35
diet, 12–14, 16
diurnal animals, 21

energy, 12–14, 16
erosion, 33

fish. *See names of specific fish*

47

fishing industry
  benefits from, 33
  cyanide and, 38–39
  damage from boats of, 34
  overfishing by, 37–38
Florida Keys National Marine
  Sanctuary, 39
fringe reefs, 16–17

gas deposits, 9
giant clams, 30
global warming, 36–37
Great Barrier Reef, 19–20

hermit crabs, 25

Indonesia, 37

jellyfish, 11

lagoons, 19
limestone cups, 14
lobsters, 26

Malaysia, 37
Maldive Islands, 18
mandarin fish, 24
marine animals, 21–22
  see also names of specific marine animals
medicines, 9
Mozambique, 39

New Guinea, 30
nocturnal animals, 22

petroleum deposits, 9
Philippines, 37, 39
photosynthesis, 8, 13–14
pink tube sponges, 26

polarized light, 23
pollution, 35–37
predators, 22

reef lobsters, 26
reproduction, 14

sea anemones, 5–6, 11
sea cucumbers, 26, 28–29
sea fans, 16
sea stars, 26–28
sea urchins, 29
seaweed, 7, 8
sea whips, 16
sharks, 22
soft corals, 16
Southeast Asia, 30
spicules, 16
sponges, 26
starfish, 26–28
stony corals
  bodies of, 8
  diet of, 13–14
  symbiosis, 5–6

tentacles, 11–12
tourism, 33–35
trash, 35
turtle weed, 7

ultraviolet light, 23–24
United States, 39–41

Vietnam, 37
Virgin Islands, 33–34

water temperature, 14, 37

zooplankton, 12–13, 14, 16
zooxanthellae, 13–14, 35